THE VERDE VAI
A Geological Hist

by
Wayne Ranney

Museum of Northern Arizona

Top: View of the Deception rhyolite in Deception Gulch. Originally erupted as a volcanic dome beneath sea level about 1.8 billion years ago.
Bottom: Close up of the Deception rhyolite. Photographs by Christa Sadler
Opposite page: View of the Precambrian rocks from within the United Verde open pit mine, Jerome. These highly deformed, ore-bearing rocks originated in a submarine, volcanic setting. Photograph by Wayne Ranney

The Precambrian

The geological record of the Verde Valley begins with rocks formed in the Precambrian Era, a time period so vast and far removed from the present that even geologists have some difficulty comprehending the length of time that it represents. The oldest rocks in the Verde Valley, exposed near Jerome, are about 1.82 *billion* years old or, stated another way, 1,820 million years old. It is almost impossible to comprehend what these numbers really mean. Most of us routinely use and speak of the number one million even though that number itself is quite large. We often hear of millionaires and know that most of us would feel quite wealthy with only a fraction of that amount. But a billion! What is a billion?

Some examples may help to illustrate the vastness of geological time. From the top of this page to the bottom, 280 sand grains would fit side by side (each grain is 1 millimeter in diameter). That means that about 1,600,000 sand grains would stretch out one mile, and 160 million sand grains would cover over 100 miles. A billion tiny sand grains side by side would stretch over 615 miles—from the City of Flagstaff to beyond San Francisco, California!

With respect to time, a billion seconds is a time period greater than 32 years. A billion minutes would take us back almost to the time of Christ, about 1,950 years. A billion years takes us back to a time when blue-green algae was the most advanced form of life. Perhaps now we can appreciate the idea of a billion and the magic contained within the antiquity of the rocks.

1.82 billion years ago, the area now occupied by the Verde Valley was unrecognizable as we know it today. What now is Arizona was situated near the edge of the ancient North American continent. Recent discoveries from other parts of the state have shown that small blocks of land, or microcontinents, moving in from the south collided with North America in a process known as continental accretion. All of the earth's continents originally were assembled in this way, with the movement of crust or plates over the earth's surface. The ancient suture lines where the microcontinents became attached to North America are only now being recognized by geologists in Arizona.

Many of the microcontinents that collided with North America in Arizona were formed in the same way that Japan and the Philippines have formed in the recent geological past. These features, known as island arcs, form when two oceanic plates collide, causing one to be submerged beneath the other. This melts the lower, subducted plate, and volcanoes will form on the surface above the melting rocks. If you envision the present setting of Japan, with respect to its position with Asia, it may be easier to understand what the area around the Verde Valley looked like 1.82 billion years ago.

The oldest rocks in the valley are layered basalt and rhyolite, rock types commonly found in island arc settings. (In discussing the early geological story of the Verde Valley, we will focus on the Jerome area because the best exposures occur here.

It also is important to remember that the present interpretation has been used only since 1971. Before that, during the heyday of mining activity, geologists believed that the environmental setting and the subsequent ore formation were vastly different. Although the rocks themselves have not changed, our understanding of how the earth operates has.) These ancient volcanic rocks are easily observed in red and brown spires of rock southwest of Jerome where Highway 89A goes up Deception Gulch. A morning drive up this stretch of winding roadway is a sight not to be forgotten!

Within the basalt are "pillow structures"—so named because these cauliflower-shaped masses of hardened lava look like pillows. These form when hot magma is erupted and then cooled rapidly in an aqueous environment. Modern pillow structures form in places like the Hawaiian Islands, and it is there that they first were recognized in the 1960s. We are fortunate that these pillow structures were preserved near Jerome because they are evidence that the volcanic rocks were erupted in a submarine setting. The basalt is called the Shea basalt (after the Shea Mine south of Jerome), and the interbedded rhyolites are called the Deception rhyolite (after the gulch). The strange, colorful spires of rock mentioned earlier are exposures of the Deception rhyolite.

Although these two formations were deposited simultaneously, they were erupted from different vents and are composed of very different kinds of volcanic rock. The Shea basalt erupted from a submarine vent about five miles south of Jerome, and the Deception rhyolite erupted very near where the town is located. Alternate eruptions caused the layers to become interbedded.

Continued eruptions from the Deception rhyolite vent eventually formed a broad volcanic dome on the ocean floor that was as much as 2,000 feet high. After the dome was built to this height, a break in volcanic activity commenced. We know this because we find thin layers of silica called chert, which could only have formed during a long period of relative quiet on the ocean floor. This chert is found in thin layers throughout the area.

Hot springs, derived from the magma below, continued to come up along fractures in the volcanic dome during this volcanic hiatus. These hot solutions contained small amounts of copper, gold, and silver in liquid form, and the tremendous drop in temperature and pressure from within the dome to the ocean floor is what caused these minerals to precipitate out onto the sea floor.

Above and right: Views of Jerome during the heyday of mining. Photographs courtesy of the Jerome Historical Society

Headframes and buildings from mining days in the Jerome open pit. Photograph by Tom Brownold

This process created the mineral-laden crusts that formed around the vents of the hot springs on top of the volcanic dome (see diagram on page 7). The Verde Central Mine, which can be seen as the gray tailings about one mile west of Jerome on Highway 89A, was developed in ore that was precipitated on this surface of the ocean floor long ago in the Precambrian Era.

A catastrophic set of events radically changed this environment after the Verde Central ore body and related cherts were deposited. Magma, perhaps mobilized when a microcontinent collided with the continental edge, came up from below and caused the center of the volcanic dome first to swell like a blister and then explode. Bits of rock and magma containing many quartz crystals were spread violently across the sea floor in sheets. Geologists call this rock the Cleopatra crystal tuff. (Before 1971, geologists believed that this rock unit formed within the crust as an intrusion, and they initially called it the Cleopatra quartz porphory.)

Continued eruption of the Cleopatra crystal tuff formed a void in the magma chamber, causing the top of the dome to collapse in an arcuate fashion along faults around its periphery. With each eruption, the dome collapsed more and more in its center. This formed what is called a cauldron (Latin, for boiling pot) on the rhyolite dome. The Cleopatra crystal tuff accumulated in great thicknesses within the subsiding cauldron and as thin sheets outside of it (see diagram on page 7). At Jerome, which would have been at the southern margin of the cauldron, the Cleopatra crystal tuff is at least 2000 feet thick. Drilling suggests that the original diameter of the Cleopatra crystal tuff may have been as much as eight miles—with a volume of six cubic miles. Some exploration geologists suspect that the northern portion of the cauldron and dome still exists beneath the Verde Valley northeast of Jerome.

It was this incredible "heat machine" of the Cleopatra crystal tuff that produced the United Verde ore deposit where the open pit at Jerome is located now. These massive sulfides formed from processes identical to those described earlier for the Verde Central deposit. This time, however, hot solutions from the magma chamber below found their way up to the ocean floor along the faults that had created the cauldron. As these solutions moved upward, they could not hold the copper, gold, and silver in liquid form because of the tremendous drop in temperature and pressure between the inside of the magma chamber and the

Above: Deception rhyolite as exposed in its type section—Deception Gulch.
Right: Quartz veins shooting through the Deception rhyolite.
Photographs by Christa Salder

ocean floor. Sulfide-rich crusts formed where these hydrothermal solutions were vented onto the sea floor at the mouths of the hot springs. The lower portions of the vents (along the faults) produced the richest ore because continued venting of the mineral waters concentrated the ore at those localities. The tops of the sulfide mounds contained massive amounts of uneconomical pyrite metal.

This fantastic concentration of ore may have occurred rather quickly, geologically speaking,

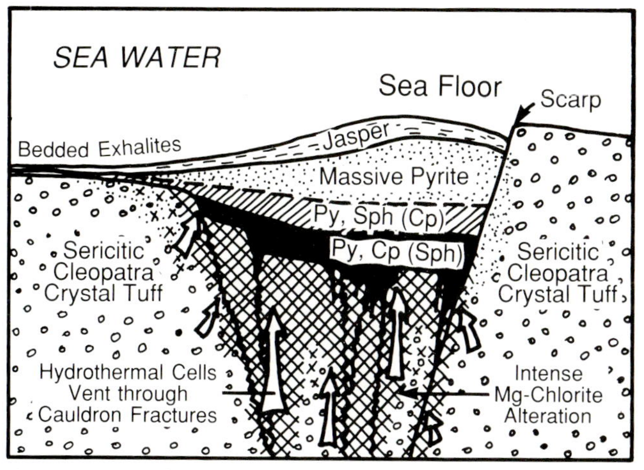

Cross-section showing hydrothermal vents (arrows) and overlying rhyolite dome. Solid black depicts zones of high mineral concentrations. (Adapted from Lindberg, 1986)

because rhyolite domes that contain no ore whatsoever immediately overlie the ore body. These rocks are called the Upper Succession rhyolite, and they are characterized by a complete lack of ore that tells us that the mineral-laden hot springs ceased to exist by the time of the Upper Succession eruption. Conversely, the Shea basalt and the Deception rhyolite are full of ore wherever they are cut by the cauldron-forming faults that vented the hot solutions. Geologists use this information to predict where future ore can be found. In places where the Upper Succession rhyolite is exposed, for example, a geologist would know to look elsewhere to locate the underlying rocks likely to contain copper ore.

The geological story of Jerome did not end with the cessation of volcanic activity. Overlying the Upper Succession rhyolite is the Grapevine Gulch Formation, a sedimentary unit that provides us with additional evidence of an oceanic

The historic town of Jerome is built directly along the trace of the Verde Fault. Foreground shows the Redwall Limestone that was dropped about 6,000 feet below Cleopatra Hill (reddish rocks behind the town).
Photograph by Wayne Ranney

environment. The Grapevine Gulch Formation is composed of material that originated in turbidity currents, submarine currents that travel at great speeds for long distances along the ocean floor.

Turbidity currents were discovered in the 1930s when the Trans-Atlantic cable was cut mysteriously by what later was described as turbidity flows. These currents carry rocks and debris suspended above the sea floor. Thus, when the material (called turbidites) finally comes to rest, it looks different than sediment deposited in a river, beach, or other environment.

The Precambrian turbidites near Jerome can be seen in Hull Canyon, where the ruins of the Walnut Springs swimming pool are located. They are composed of greenish layers of silt and angular volcanic debris called breccia. These turbidites probably originated when earthquakes dislodged the upper portions of the volcanic dome and the resulting landslide created a turbidity current. Layers of red chert are found interbedded within the Grapevine Gulch Formation, suggesting that long periods of time may have elapsed between the deposition of various turbidites.

Pillow lavas over 1.8 billion years old, Shea basalt, Jerome.
Photograph by Christa Sadler

The final rock unit of this great pile at Jerome is an intrusive layer that never reached the surface of the ocean floor. It is called a gabbro, a rock type identical to basalt except that it cools slowly beneath the surface of the earth and thus has time to form crystals. The gabbro can be seen in the northwest wall of the open pit.

With the emplacement of the gabbro within the old volcanic dome, the depositional and eruptive record of the Verde Valley's Precambrian history was complete. What happened next relates directly to the accretion of the North American continent. Collision and accretion of other land masses from the south caused intense compression of the crust, and the entire volcanic pile was deformed by folding and stretching of the rock units. Amplitudes of some of the folds are up to 6000 feet, suggesting a collision of major proportions.

The first collision, which occurred about 1.74 billion years ago, wrinkled the rocks at Jerome into these spectacular folds. It is no wonder that our interpretation for how these rocks originally formed has changed so radically in the last few years. These volcanic rocks were subjected to a head-on collision of great intensity, and this crumpling and folding of the original pile is what frustrated early miners in their attempts to locate the valuable ore. This deformational event caused the entire area to be uplifted. Erosion then worked vigorously to wear down the uplift during the next few 100 million years.

Indeed, as we shall see later, it is incredibly fortuitous that the copper ores at Jerome were not eroded away at various times in the geological past or buried beneath younger rocks. What are the chances that other ore-rich cauldrons are present beneath Clarkdale, Cottonwood, or Sedona? In the short run, it probably does not matter because exploration for buried ores is difficult, and extraction is very uneconomical. Still, it makes us wonder—what is under all of those rocks?

The Early Paleozoic

Whereas the Precambrian rocks exposed at Jerome give evidence for a submarine volcanic setting, most of the Paleozoic rocks in the Verde Valley record a time of hot, desert environments. It may come as a surprise to the uninitiated that an area as compact as the Verde Valley should change its character so dramatically. But, again, it is the vast span of geological time that accounts for this change. The oldest Paleozoic rocks found in the Verde Valley are 1.2 *billion* years younger (1,200 million!) than the Precambrian rocks at Jerome. Given this amount of time, dozens of separate and distinct environments may have been present in the area of the Verde Valley. Understanding the immensity of geological time is the fundamental tool necessary for anyone to become an earth scientist.

Top: Martin Formation and a sycamore tree along the Verde River near Childs.
Bottom: Chert nodules within the Martin Formation in the Jerome area. Photographs by Wayne Ranney

How can we know anything about what the Verde Valley was like during this 1.2-billion-year period for which there is no rock record? It is almost impossible since we have no solid evidence. (These gaps in the rock record are known as unconformities, and the particular one we are speaking of here is called The Great Unconformity because of the length of time it represents.) However, geologists can look at nearby areas where rocks have not been eroded completely within The Great Unconformity and obtain some information about this period of time. In the Grand Canyon and central Arizona, for example, there are sedimentary rocks known respectively as the Grand Canyon Supergroup and the Apache Group that were deposited at this time. It is a fascinating sequence of rocks that allows the observer to glimpse a very ancient landscape of Arizona.

In the Grand Canyon, almost two and one-half miles of layered sediment and lava accumulated in a 350-million-year period from 1,200 to 850 million years ago. Most of these deposits were laid down in shallow marine or coastal settings before any type of life was present on land. About one billion years ago, lava flows were erupted on a coastal plain in the Grand Canyon area, and it is easy to imagine the black lava flows being buried by the mud of ancient tides. The fissures that released the lava are visible today within the Grand Canyon at Hance Rapids and in central Arizona where Highway 60 crosses the Salt River Canyon. Because the Verde Valley is located between these two areas, we can infer that these environments were present there as well.

At about 850 million years ago (mya), deposition was halted when western North America again was involved in some type of collision or rifting event. This event caused the two and one-half miles of layered sediment and lava to be tilted and faulted into alternating high and low blocks. Erosion removed all of the high standing blocks and most of the lower ones. Only very low blocks were preserved from the ensuing 300 million years of erosion. The fortuitous preservation of these low blocks occurred only in the Grand Canyon and in central Arizona, where the Grand Canyon Supergroup and the Apache Group rocks are found.

In the Verde Valley, none of these sediments survived this period of erosion, which suggests two possibilities—either the blocks of sediment were not faulted as low, or erosion was greater. We are fortunate that the deformed volcanic dome of copper at Jerome was not eroded also at this time. If this erosional period had lasted perhaps only five or ten million years longer, we may never have seen the birth of Jerome as one of Arizona's premier mining towns. It was the next cycle of deposition that preserved the ore bodies from their near miss with erosion.

The first period of the Paleozoic Era is called the Cambrian. The word comes from the Roman name for Wales (Cambria), where rocks of this age, about 570 to 500 million years old, first were described. The most complete Cambrian deposits in the world, however, are in the Great Basin of eastern California, Nevada, and western Utah. During the Cambrian, this area subsided into an ocean basin that received sediment without interruption. For this reason, there are very few unconformities within the Cambrian section of rocks in the Great Basin. Farther to the east, in the Grand Canyon and the Verde Valley areas, subsidence of the crust was not as great, and only a thin veneer of sediment accumulated in these regions.

Above: Blocks of Tapeats Sandstone were quarried for many of the mining era buildings in Jerome.
Opposite page: Ruins of the Bartlett Hotel in Jerome. Archways were cut from the Tapeats Sandstone, a nearshore beach deposit over 500 million years old.
Photographs by Wayne Ranney

The first deposits, known as the Tapeats Sandstone, accumulated in a beach environment along a shoreline that stretched from Sonora, Mexico, to British Columbia, Canada. North America was not colliding with any other land mass at this time, so the west coast of the continent (which would have been the north coast in the Cambrian) was a low-lying, featureless plain in which sediments of similar composition accumulated. The Flathead Sandstone in Montana and Wyoming is identical in age and appearance to the Tapeats; the difference in names results from geologists working in widely divergent areas earlier in this century, and the old names remain. Many of the old, historic buildings in Jerome are constructed of blocks of this brown-to-purplish sandstone. The next time your friends from out of state say that all Arizona lacks is a beach, you can take them to Jerome, where exposures of this beach deposit are common.

The Martin Formation was laid down about 165 million years after the Tapeats. This is an unconformity of considerable length, yet it is the shortest one we have discussed yet. In Nevada and western Utah, there are rocks that were deposited at this time, but Arizona was above sea level and unable to collect and preserve sediments from these time periods. By the Devonian Period (named after exposures at Devonshire, England), Arizona was below sea level, and sediments of the Martin Formation were accumulating. The first land plants appeared on planet Earth immediately prior to the dawn of the Devonian, frequently called the Age of Fishes because of the radiation of fish in the world's oceans.

The Martin Formation, which is about 800 feet thick in the Verde Valley, is best observed in the canyon between Cottonwood and Jerome along Highway 89A. It is composed of gray limestone and dolomite with many fossils of brachiopods, crinoids, sponges, and bryozoans. Some of the layers have accumulations of organic material that smells like petroleum when struck by another rock. Although not found in sufficient concentrations for oil production, these thin lenses of fetid dolomite have resulted in millions of dollars being spent looking for a possible oil reservoir in the central Verde Valley. Rocks of the same age and composition in North Dakota and Saskatchewan have yielded incredible volumes of the black gold.

The Redwall Limestone was deposited across Arizona in a shallow marine setting on top of the Martin Formation. The Redwall, like the Tapeats, can be found throughout western North America and goes by different names because it is so extensive. In southern Arizona, it is called the Escabrosa Limestone; in Nevada, it's the Monte Cristo Limestone; and in Montana, the Madison Limestone. All of the deposits reveal that during the Mississippian Period (about 350 mya), this part of North America was a shallow, tropical, coral sea that lay very near the equator. The best place to observe the Redwall is near the top of Mingus Mountain on Highway 89A.

Redwall Limestone and prickly pear cactus. Originating in a tropical sea about 350 million years ago, the Redwall is quarried extensively for cement in the Verde Valley. Photograph by Christa Sadler

The Late Paleozoic

The old "Supai Formation" of McKee now is classified as the Supai Group (lower blocky cliffs), the Hermit Formation (middle, tree-covered slopes), and the Schnebly Hill Formation (upper orange spires and cliffs). Photograph Ralph Hopkins

Without a doubt, the most spectacular rocks within the Verde Valley are the red rocks exposed near Sedona that were deposited during the Pennsylvanian and Permian periods. It was an exciting time in North American geology because after a 375-million-year journey to the east, our continent finally collided with Europe, Africa, and South America to form the giant supercontinent of Pangaea. This collision, which is similar to the one occurring today between Asia and India, was the final pulse in activity that created the Appalachian Mountains in North America and the Caledonean Highlands in Great Britain.

This lofty range was mantled on its lower slopes by the thick, tropical vegetation that later became the extensive coal deposits of Pennsylvania (thus, the name of the time period), Ohio, and West Virginia. This was on the west side of the mountains. On the east side, similar tropical vegetation has become coal deposits in France, Germany, and Poland. Reptiles evolved at this time,

Vultee Arch eroded out from the Schnebly Hill Formation in the Secret Mountain Red Rock Wilderness Area. Photograph by Tom Brownold

Midgley Bridge on U.S. Highway 89A is built upon the Supai Group (light-colored rocks). Tree-covered slope behind the bridge belongs to the Hermit Formation, and dark spires above are in the Schnebly Hill Formation. Photograph by Tom Brownold

perhaps from a common ancestor to amphibians, and precursors to the dinosaurs roamed the plains of Oklahoma, Texas, and New Mexico.

The area of the Verde Valley lay at the extreme western edge of the giant supercontinent, and it may have been possible to travel overland through the reptile flats in Oklahoma, over the snowy summits of the Appalachians, and onto the continents of Africa, India, Australia, and Antarctica, all without having to cross a body of water larger than a river!

The names of the various formations making up the red rocks around Sedona have been the source of much confusion to residents, visitors, and even geologists. Various workers have tried to relate the deposits at Sedona with those at the Grand Canyon, and although they are identical in most respects, there are enough differences to cause controversy. In 1945, Edwin D. McKee, the preeminent authority on the geology of the Grand Canyon and the southern Colorado Plateau, published in *Plateau* (Volume 18) a classification to the rocks in the Oak Creek area. He proposed that all red rocks in the area be assigned to the Supai Formation with three informal subdivisions called the "A," "B," and "C" members. This classification worked well and followed what was observed at Grand Canyon, except for two things: 1) the Supai Formation in Oak Creek was much thicker and was composed of more sandstone than the rocks at the Grand Canyon; and 2) the Hermit Formation was not recognized in Oak Creek.

In 1979, Professor Ronald Blakey of Northern Arizona University began to look closely at the red rock cliffs along Schnebly Hill Road. Other geologists—Don Elston, of the U.S. Geological Survey in Flagstaff, and Wes Peirce, of the Arizona Geological Survey in Tucson—also have attempted revisions of McKee's original classification, but no firm agreement among the geologists has been

Conglomerate deposit from within the Hermit Formation. These clasts were derived in river systems that originated in the Ancestral Rocky Mountains about 270 million years ago. Photograph by Christa Sadler

reached. In science, new ideas are never accepted immediately, and some geologists still propose names for rocks near Sedona that are different than those presented here. But more and more, the classification first proposed by Ron Blakey in 1979 is becoming the red rock language of the Verde Valley.

This classification correlates the "C" member of McKee with the true Supai Formation; the "B" member now is called the Hermit Formation; and the "A" member is a newly described unit not found in the Grand Canyon, called the Schnebly Hill Formation. (Note: in 1975, Eddie McKee further divided the Supai Formation into four separate and distinct formations and proposed the name Supai Group for the entire package. Don't be confused—only the term Supai Group will be used here when referring to the old "C" member of the Supai Formation.)

Beginning about 300 million years ago, the area of the Verde Valley that for so long had been covered by warm, tropical seas began to emerge. The lowermost portions of the Supai Group show alternating layers of gray, marine limestone (from the receding tropical seas) and red mud and sand deposited in coastal deltas that covered the marine deposits. Within the Supai Group, from bottom to

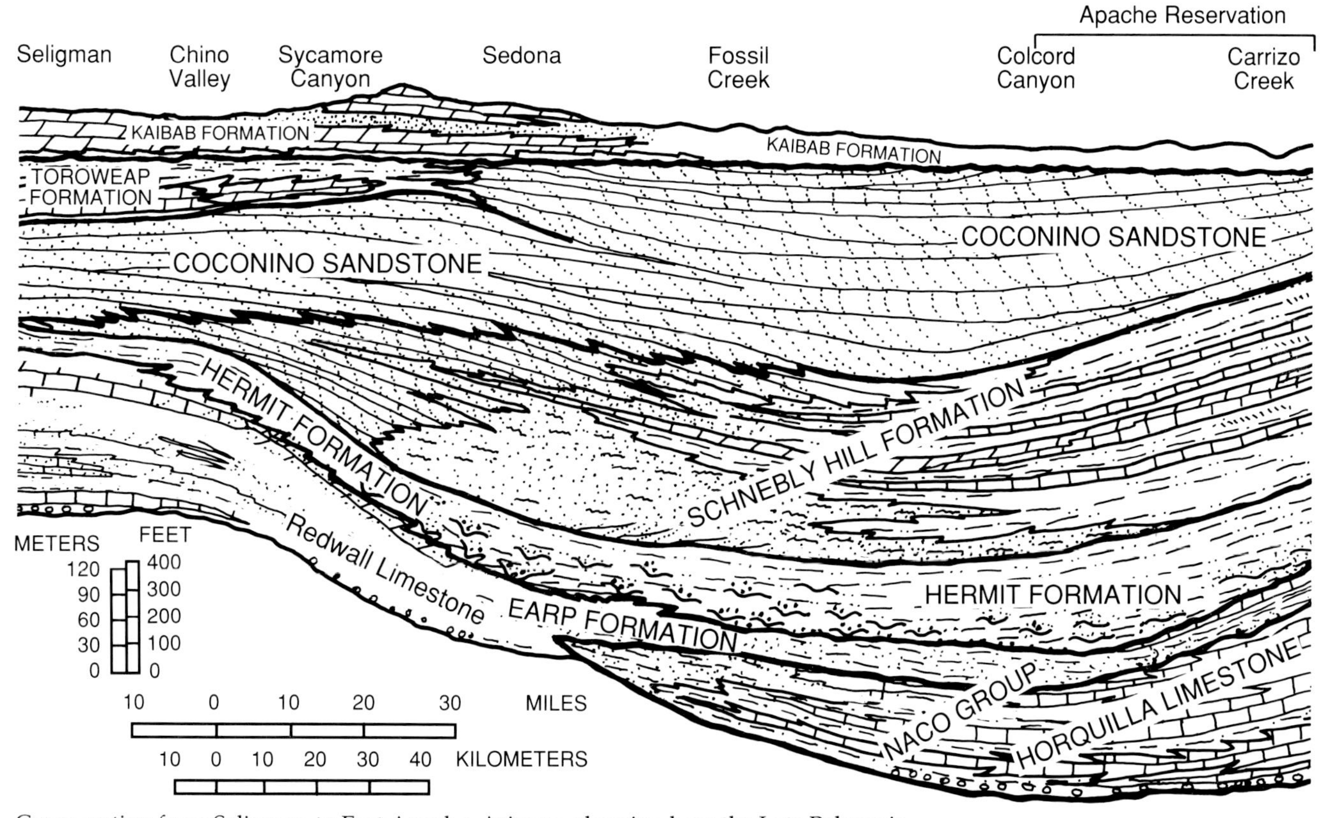

Cross-section from Seligman to Fort Apache, Arizona, showing how the Late Paleozoic rock units thicken and thin in this part of the state. (Adapted from Blakey)

Opposite page: Cliffs of sandstone deposited in ancient desert dunes make up the walls of the West Fork Canyon.
Left: Ferns and cross-bedded sandstone.
Bottom: Basalt boulders along Oak Creek. Photographs by Ralph Hopkins

top, we can observe how the climate, topography, and landscape of this part of Pangaea changed.

It is as if there were a battle at this time between shallow marine seas and coastal deserts of wind-blown sand. It takes a trained geologist working with a microscope and rock thin-sections cut to 3/1000's of an inch to observe the remains of this battle between the sea and the desert. Polarized and unpolarized light is projected through thin sheets of rock to see how the sediments have been altered after deposition. What we see is that much of the limestone and quartz within the Supai Group has been replaced by dolomite, which is evidence that the sea advanced and retreated many times over the coastal lowlands. However, it does not always take a trained eye working with a microscope to witness the results of this struggle. The best place to observe it is beneath Midgley Bridge about one mile north of Sedona on Highway 89A. At the bottom of the canyon, gray fossiliferous limestones deposited in a shallow sea give way upwards to the deltaic, fluvial, and/or wind-blown sands found directly beneath the bridge. It's a difficult climb down, but you can observe the changing nature of the earth from the parking area at the bridge.

The next formation is one that many Sedonans should be familiar with, whether they know it or not. It is the Hermit Formation (the old Supai "B" member), and it is the rock unit that most of the city is built on. It erodes rather quickly, geologically speaking, and has formed the broad, rolling platform of Grasshopper Flat (West Sedona). Harder formations form cliffs and spires of rock; softer ones like the Hermit Formation form gentle slopes of eroded debris.

The Hermit Formation is an interesting one because it and its equivalent deposits in Utah and Colorado reveal a detailed record of the geography of the Four Corners region about 270 million years ago. Immediately prior to this time, a mountain range was uplifted between Green River, Utah, and Taos, New Mexico, through southwestern Colorado. Because the ancient range was located where the modern Rocky Mountains are, it is referred to as the Ancestral Rockies. Except for its location, the Ancestral Rockies were not connected in any way with the modern mountains. These old mountains, according to the most recent interpretation, were the result of the collision between North and South America during the assemblage of Pangaea. Similar ranges formed in a southeast-directed arc in line with the Ancestral

Rockies in Oklahoma, Texas, and Arkansas at the same time.

It was sediments shed off of the Ancestral Rockies, however, that are important to our story. Near Moab, Utah, at a place called Fisher Towers, there are exposures of red sediment that contain cobbles and boulders of Precambrian igneous rocks. It is unusual to preserve sediments with such large clasts as boulders, but preserved they were as the remains of the alluvial fans that were aggrading at the base of the Ancestral Rockies. If you follow these deposits to the southwest, you notice that the boulders and cobbles disappear gradually as sand and mud become the dominant sediments. The same is true of modern mountain fronts; the big materials fall out of the rivers closer to the mountain front, and the sand and mud are carried farther towards the sea.

It is possible to follow the 270-million-year-old river system of the Ancestral Rockies from near Moab, through Canyonlands National Park, into

Monument Valley (where the equivalent Organ Rock Formation makes up the sloped bases of the monuments). It continues on into the subsurface beneath the Hopi Mesas and Flagstaff and finally reaches the Mogollon Rim and Sedona. From evidence of its path, we are able to determine that the Hermit Formation originated in a floodplain from rivers that drained the Ancestral Rockies in Colorado 270 million years ago.

One of the best places to see evidence for the fluvial origin of the Hermit Formation is at the upper parking lot for Grasshopper Point, adjacent to Highway 89A. In the roadcut to the west of the highway, we can observe channels that were cut into the mudflats and filled with gravel. The mudstone was deposited on a floodplain—defined as the area outside of the river channel that is covered with water only during large floods. The fluvial system cut an arroyo, or gully, into this floodplain deposit and gradually filled it with gravel. At the Grasshopper Point upper parking area, we also can observe that additional channels were cut into the gravel deposit and later filled with floodplain deposits, evidence that the area underwent alternating periods of cutting and filling. These cut and fill relationships, in addition to caliche deposits, suggest that the climate near Sedona in the early Permian was arid and subject to the flash flooding that is common in our arid environment today.

It is possible to visualize the entire geographical setting from the arid mountain front in Colorado to the dry coastal plain near Sedona. A good modern analog may be in Iran, where the Elburz Mountains north of Tehran are being uplifted because of the collision of the Arabian plate with Asia. The rivers on the south side of this range dry up in the desert flats that stretch to the south towards the Persian Gulf—endless cycles of time, climate, and landscapes.

With the passing of time, the environment slowly changed in the area of the Verde Valley. Sand, first in isolated patches and later in colorful dunes, began to blow across the Hermit floodplain

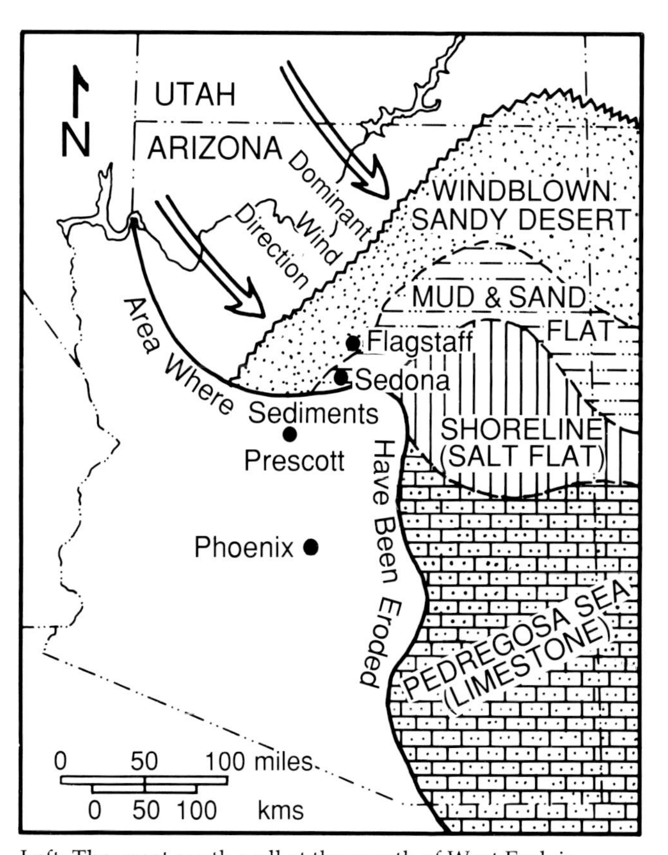

Left: The great south wall at the mouth of West Fork is composed of cross-bedded Schnebly Hill Formation and Coconino Sandstone. Photograph by Tom Brownold
Above: Map of Arizona about 270 million years ago. Wind from the northwest deposited sand in the Sedona area as the Schnebly Hill Formation. (Adapted from Blakey)

from the north. This sand was derived from the erosion of older rocks, perhaps from as far away as Canada. We know that the wind blew the sand in from the north because of the way in which it is preserved. The evidence, called cross bedding, is nothing more than the preserved portion of the lee (or downwind side) of a sand dune. Geologists have determined that cross bedding represents the dominant wind direction by cutting into modern sand dunes. More than eighty percent of the crossbeds in the Sedona area dip to the south or southeast. Crossbeds dipping in other directions formed from blowing sand that swirled around the edges of large dunes. These sand deposits, which eventually buried the entire Hermit floodplain, are called the Schnebly Hill Formation (the old Supai "A" member). This formation makes up all of the colorful buttes and spires of rock in the Sedona area—such as Coffee Pot Rock, Cathedral Rock, and Bell Rock.

Beautiful swirls of quartz sand in the Coconino Sandstone represent the desert dune conditions that were present in the area of northern Arizona. Photograph by Christa Sadler

The flat-bedded Kaibab Formation represents marine deposits that covered northern Arizona at the end of the Paleozoic era. This resistant cap rock "holds up" much of the Mogollon Rim and the Grand Canyon. Photograph by Ralph Hopkins

Basalt Flow in Fossil Creek, one of the Mogollon Rim's beautiful drainages. Photograph by Stephen Trimble

The Schnebly Hill Formation is another unit that reveals remarkable detail of its depositional environment. From the bottom to the top of the formation, we are able to document how the desert dunes evolved through time. We are able to tell, for example, that the first sand dunes coming in from the north traveled across the Hermit mud flats where the Grand Canyon now is located. Since this area was slightly higher than the Sedona area, the Schnebly Hill sands were transported across that region. That is why there is no Schnebly Hill Formation in the Grand Canyon.

Merry-Go-Round, type section of the Schnebly Hill Formation. Photograph by Wayne Ranney

Sand dunes blowing into the south Atlantic Ocean in Namibia may be the best modern analog for how the Sedona area looked about 167 million years ago. Photograph courtesy of USGS

Around Sedona, and farther to the east, gentle subsidence was occurring. Thus, as sheets of sand came in from the north, they were lowered and buried continually by new sand dunes. Because of this subsidence, a shallow marine basin encroached on the Sedona area from the southeast. It is referred to as the Pedregosa Sea. As the Schnebly Hill sands were blown into the subsiding area, they were reworked within the broad tidal zone of the Pedregosa Sea. For this reason, the lower sections of the Schnebly Hill Formation are flat bedded rather than cross bedded. Whenever you are in the area, look at the lower portions of the colorful red rock spires and see if you can spot this reworked sand.

Approximate 267 million years ago, the subsidence of the basin was great enough that the Pedregosa Sea covered the Schnebly Hill sands at Sedona. Another possibility is that the sand supply from the north was temporarily cut off. This event is recorded by the Fort Apache Limestone, a thin, gray marine unit that is present about two-thirds of the way up from the bottom of the 700 feet of Schnebly Hill Formation. The Fort Apache limestone is only six feet thick at Sedona, whereas, at Fort Apache to the southeast, it is over 100 feet thick. This is evidence that the Pedregosa Sea was centered to the southeast and intruded into the Sedona area for only a short geological instant. In fact, in Boynton Canyon northwest of Sedona, the Fort Apache Limestone pinches out entirely, documenting the very edge of that Permian shoreline.

As the Pedregosa Sea finally retreated for the last time to the southeast, the Sedona area was buried again by large quantities of windblown sand. Huge dunes traveled across the Fort Apache Limestone surface, burying it with highly cross-bedded sand. The Schnebly Hill Formation above the Fort Apache Limestone is almost entirely cross bedded, which tells us that these dunes were not reworked by the tides. You can see these petrified swirls of sand along Schnebly Hill Road at Merry-Go-Round. A modern analog of this unique desert setting may be found in Namibia (southwest Africa), where great quantities of red, windblown sand are found in dunes adjacent to the south Atlantic Ocean.

Through time, the coastal dunes of the Schnebly Hill Formation pushed farther to the southeast, where they can be seen in walls of upper Fossil Canyon. The dunes that followed these were, by definition, inland dunes since they were farther inland from the shoreline. This gradual change resulted in the gradational contact of the Schnebly Hill Formation with the overlying Coconino Sandstone. It is difficult to determine the precise point the Schnebly Hill Formation ends and the Coconino Sandstone begins. Generally, we can say that contact occurs at the highest red sandstone. Remember, however, that many times geologists try to draw lines where there are no lines. The change from coastal dunes to inland dunes in the Sedona area came about because of the great quantities of sand that pushed the shoreline farther to the southeast. This period of change was characterized by continuous deposition, and you can think of this relationship as the opposite of an unconformity (there is little or no gap in the rock record).

Within the Coconino Sandstone are lizard tracks and raindrop impressions that are the key indicators of this inland dune environment. These, and the ubiquitous south-facing cross beds, testify to the lack of surface water present in this area at this time. With deposition of the Coconino Sandstone at about 265 million years ago, we witness the climax of 85 million years of gradual environmental change in the Verde Valley area—from the tropical seas of the Redwall Limestone 350 million years ago to the desert dunes of the Coconino. Each of these gradual changes in climate and environment is recorded beautifully in colorful rock exposures that are easily viewed and enjoyed. Ours is the first species, and we the first generation of that species, to be able to look at the rocks and comprehend what they have recorded from these long-lost scenes of earth history.

Two additional Permian age formations overlie the Coconino Sandstone: the Toroweap and the Kaibab formations. Both are marine deposits that originated from the west. There is some confusion as to whether the name Toroweap should be used in the Oak Creek area. The trouble is caused by its close resemblance to the Coconino Sandstone east of Oak Creek Canyon. If you have visited Walnut Canyon National Monument east of Flagstaff, you have seen this sandy version of the Toroweap.

At Sycamore Canyon and in most of the Grand Canyon, the Toroweap is a silty sandstone with large quantities of gypsum. It almost always forms slopes of easily eroded debris. However, within a five-mile-wide zone that trends from the Verde Valley to Glen Canyon Dam, the Toroweap gradually changes from a softer, slope-forming unit into a cliff-forming sandstone that looks very much like the underlying Coconino Sandstone. This lat-

Fossil trackways of lizard, Coconino Sandstone. Photograph by Stephen Trimble

eral change of rock type within a formation is called a facies change. It records the differences in environment from one locality to the next during the same time period.

The facies change that occurs within the Toroweap Formation from Sycamore Canyon to Oak Creek Canyon records an evaporating, shallow-marine mud flat to the west and a sandy, near-shore beach enviroment just five miles east at approximately the same time. A trained observer can trace this change of rock type from Sycamore to Oak Creek. Even when observed in Oak Creek Canyon (where the Coconino and Toroweap are nearly identical), differences between the two are discernible. For this reason, the Toroweap terminology is used here. However, if you come across literature or lecturing geologists who fail to mention the Toroweap Formation in Oak Creek Canyon, remember that it is a deposit undergoing a facies change and that what you call it does not change what it is.

The Kaibab Formation is the final chapter of our Paleozoic story. It is a limestone, dolomite, and chert deposit that caps most of the Mogollon Rim country and the Grand Canyon. In fact, it is the Kaibab's resistance to erosion that has allowed the preservation of much of the previous rock record. Were it not for the silicious chert layers within the Kaibab, much of the previously discussed history may have been eroded before our generation of "rock readers" could decipher the codes.

Studies at the Grand Canyon suggest that the chert layers in the Kaibab Formation originated from sponges that grew on the sea floor about 260 million years ago. This came about because many species of sponges have spicules in their bodies. Spicules are thin, needlelike rods of silica, and when sponges die, millions of these silica needles fall to the ocean floor. Through time, billions of these spicules will form sheets of silica. Later, this can be compressed into a dense chert highly resistant to erosion. Red chert is called jasper; black chert is called flint. The chert in the Kaibab, which is cream colored, is sometimes called novaculite. Whatever you call it, it was prized by the local Sinagua culture for use in making projectile points. It is doubtful that they were aware that their sharp stones originated in sponges 260 million years ago. But believe it or not, it is those ancient blooms of sponges that have preserved the spectacular scenery of the Grand Canyon and the Mogollon Rim in Arizona!

The Mesozoic

When the Kaibab seas finally retreated about 260 million years ago, there were other environments that may have left sediment on top of the Kaibab Formation. But the following 35 million years of earth history have no record on the Colorado Plateau. Sediment from this period either was never deposited or eroded away before the Triassic Moenkopi Formation came in and buried the Kaibab about 225 million years ago.

The Moenkopi is a very red, thinly bedded sandstone that records mostly flat, near-shore tidal environments. Remnants of the Moenkopi Formation are found only on the plateau surface above Sycamore Canyon (and behind the courthouse in downtown Flagstaff). On the Navajo Indian Reservation, in Zion National Park, and in other areas of the Colorado Plateau, over 10,000 feet of additional sediment accumulated on top of the Moenkopi during the next 165 million years. This period of time, the Mesozoic Era, also is known as the Age of Dinosaurs, and in many localities on the plateau, their remains can be found weathering out of volcanic ash and river or swamp deposits.

Dinosaurs almost certainly were present in the area of the Verde Valley, but the sediments that would have held their bones have been eroded away. The next time you visit the Verde Valley, look to the sky above the Mogollon Rim and think of the Mesozoic age rocks that once were present here. Imagine the endless landscapes, similar to those seen in southern Utah today, that appeared and disappeared as this 10,000 feet of colorful sediment was slowly eroded away, never to be seen by human eyes! As earth scientists, however, we can look at the rock record elsewhere and use our imagination to bring back the scenic vistas that preceded our species.

The Cenozoic

Storm building over Oak Creek Canyon (Mogollon Rim). Photograph by Tom Brownold

How ironic that the most recent chapter of the Verde Valley's geological history, the Cenozoic, is the one that is the least understood. Although it seems that the most recent history should be the easiest to document, the Cenozoic throughout much of Arizona was a time of erosion, and rocks either were never deposited or were stripped away. Much of our knowledge of this time period has come about only recently, and the story presented here contains information that was nonexistent as recently as ten years ago. Future studies may refine this story along the way, but enough has been learned from the rocks to present the following interpretation.

The cause for the erosion of the dinosaur-bearing rocks in the Verde Valley area was uplift in the area of west-central Arizona and the modern Rocky Mountains. The last years of the Mesozoic Era saw the continued uplift of the southwestern margin of North America and the gradual retreat of the great Cretaceous Seaway to the Gulf of Mexico. The great Cretaceous Seaway bisected North America for about forty million years into two emergent land areas. The western shoreline of the seaway was in eastern Arizona and western New Mexico. Great dinosaurs roamed the coastal plains in central Utah and northern Arizona. The vegetation they thrived on was compacted later into the

Exposure of the Beavertail gravel on Highway 179. This gravel was deposited at the base of the Mogollon Rim about twenty million years ago. Photograph by Christa Sadler

coal deposits now being mined on Black Mesa, near Gallup, New Mexico, and Price, Utah. All western U.S. coal deposits formed at this time in environments very similar to the present swamplands of Virginia, South Carolina, and Georgia. The only noticeable difference is the large reptiles that inhabited the earlier swamps.

North America, which began its westward drift away from Pangaea during the deposition of the Moenkopi Formation (about 225 mya), collided with another western plate. This caused the Intermountain West to become wrinkled into alternating basins and uplifts, and the Rocky Mountains were born. This mountain-building event is called the Laramide, named after the Laramie Basin in Wyoming. The Laramide had a significant effect on the landscape of northern Arizona.

Lower member of the gravel at Beavertail Butte. Angular texture of the clasts tells geologists that these cobbles did not travel far, and their composition suggests they were derived as talus debris from the ancestral Mogollon Rim. Photograph by Wayne Ranney

During this period, and even prior to it, central and western Arizona were uplifted into a mountain range that is referred to as the Mogollon Highlands, after the escarpment they helped to create. This uplift in the central part of the state caused the sedimentary layers of the plateau region to be tilted down to the northeast towards the Four Corners region. Mesozoic and Paleozoic rocks in central Arizona formed the highest parts of the Mogollon Highlands, and this is when the dinosaur-age rocks were eroded away in the Verde Valley.

The chert layers within the Kaibab Formation provided a resistant surface of erosion, across which rivers from the Mogollon Highlands deposited gravels. Geologists refer to these deposits informally as the "Rim gravels" because of their present location on top of the Mogollon Rim. The gravels are composed of rounded Lower Paleozoic and Precambrian age pebbles that could have been derived only from central Arizona because that is where rocks of that age have been exposed. These pebbles provide evidence for the existence of the highlands even though those mountains are gone now. The "Rim gravels," too, were eroded extensively, and isolated remnants of them are the only evidence we have for what occurred in the Mogollon Rim region for at least 30 million years.

There are questions as to how old the "Rim gravels" actually are, but evidence from near Show Low, Arizona, indicates that they were deposited between 55 and 30 million years ago. Many geologists state that the Mogollon Rim could not have formed prior to deposition of the "Rim gravels" because the rim would have blocked their transportation to the northeast. Whatever explanation is put forward, there is evidence that the Mogollon Rim developed near its present location sometime between 40 and 30 million years ago.

The evidence for this is another package of gravels, which are similar to the "Rim gravels" but are located at the *foot* of the Mogollon Rim. These gravels, unnamed as yet, are found in the Verde Valley along State Highway 179, south of Oak Creek Village where Forest Road 120 turns off to Cornville. The gravels are exposed clearly at Beavertail Butte southwest of the road junction, and geologists recently have referred to the rocks as the Beavertail gravels, or Beavertail Butte formation. They are composed of two types of gravel, a lower one that contains angular Kaibab Formation and Coconino Sandstone cobbles and an upper gravel composed of rounded Lower Paleozoic and Precambrian pebbles. Both of these thin abruptly and pinch out to the northeast of Beavertail Butte against a rising slope of the Schnebly Hill Formation. Although this rising slope has been truncated by later erosion, it represents the location of the ancestral Mogollon Rim when the gravels were deposited.

The lower gravel, composed of the angular Kaibab and Coconino cobbles, represents talus debris that accumulated on the slopes of this ancient rim. The upper gravel, containing rounded (i.e., river) pebbles, could have been derived only from the southwest, where the Precambrian rocks were eroding in the Mogollon Highlands. What a setting this must have been!

Many of the basalts near Sycamore Canyon erupted during the Hickey Volcanic Event between ten and fifteen million years ago. Photograph by Stephen Trimble

Interestingly, the evidence for flow direction in the upper gravel is to the southeast, which tells us that as the rivers from the Mogollon Highlands encountered the ancestral Mogollon Rim, they were deflected in that direction. These southeast-directed gravels may be the first evidence we have for the initiation of the Verde River system into the Verde Valley.

There is good evidence that both of these gravels accumulated beneath the ancestral Mogollon Rim before the rim eroded northeastward to its present location. We suspect that the rim has been eroding or retreating in that direction since its inception. The primary evidence for this is the fact that the northeastern tilt of the rocks (which was imprinted on them when the Mogollon Highlands

Left: On top of House Mountain—a shield volcano erupted between fifteen and thirteen million years ago at the base of the ancestral Mogollon Rim.
Center: Caves weathered within the Verde Formation near the Verde River.
Bottom: View of the trace of the ancestral Mogollon Rim (far left) and the Village of Oak Creek. The rim has retreated to its present location at a rate of one foot every 625 years. Photographs by Wayne Ranney

were uplifted) causes the strata to erode down dip. The age of the Beavertail gravels—and, thus, the minimum age of the rim—is difficult to determine because no fossils have been found in them, and there are no interbedded volcanic rocks. They probably were deposited somewhere between thirty and twenty million years ago.

Whatever their age, the Beavertail gravels were partially eroded after being deposited, and the environment of the Verde Valley area was changed dramatically with the widespread eruption of basalt that began about 15 million years ago. These basalts can be observed between Prescott and Sedona and from Sycamore Canyon to Black Canyon City and New River. They are informally called the Hickey formation—named for a small mountain of the same name near Jerome. A major center of the eruptions was on Mingus Mountain, and the top of this feature exposes thick accumulations of basalt. The Black Hills had not yet been uplifted, and great quantities of lava covered the valley to the southwest of the ancestral rim.

Another eruptive center was at House Mountain, a volcano located between Sedona and Cornville, near Page Springs. It erupted 14.5 million years ago in an almost unbelieveable setting immediately beneath the ancestral rim. In fact, its crater was very near the base of the rim, and for this reason, lava could flow only to the west, the south, and the southeast. To the north and northeast, lava was ponded against the base of the rim, where erosional remnants of the Beavertail gravels were located. What a scene it must have been to look down from the top of the ancestral rim into the mouth of an erupting volcano. Of course, this would not have been possible because *Australopithicus*, an ancestor to *Homo sapiens*, was just coming out of the trees and onto the grasslands in eastern Africa.

In the last 15 to 13 million years, the Mogollon Rim has retreated to its present location four miles to the northeast of the House Mountain volcano. This has left the basalts, which were ponded against the ancestral rim, standing in erosional relief above the Village of Oak Creek. This is called inverted topography because areas that previously were low in elevation are now standing high. You can see this inverted topography and the trace of the ancestral Mogollon Rim when looking southwest from the Village of Oak Creek. Measurements made from the location of the ancestral rim to its present location (about four miles) and divided by the time elapsed since the last eruptions at House Mountain (about 13 million years) give us an average rate of retreat for the Mogollon Rim of about one foot every 625 years. At this rate, the colorful cliffs of the rim will retreat to downtown Flagstaff in another 79 million years!

Extensive volcanism related to the Hickey eruptions continued until about ten million years ago. At this time, the Oak Creek Fault was active, and an ancestral Oak Creek drainage had eroded a broad valley into the Mogollon Rim between the present day Oak Creek Canyon and Wet Beaver Creek Canyon to the east. That is why you do not see any of the colorful red rocks where Interstate 17 enters the Verde Valley from the rim. The removal of the red rocks within this broad valley allowed for the accumulation of gravels on the valley floor. These gravels can be seen across the highway from Slide Rock State Park in Oak Creek Canyon and on Horse Mesa southeast of the Village of Oak Creek. They were thought, until recently, to be part of the "Rim gravels," but detailed studies have determined that they were derived from the north and transported southward in ancestral Oak Creek.

About eight million years ago, lava from the Colorado Plateau edge covered these "Slide Rock gravels." These lava flows formed the ramp of basalt that engineers later would utilize for the placement of Interstate 17. Geologists refer informally to these flows as the "ramp basalt." One of the flows has been dated radiometrically at 6.4 million years, in an area that was most likely at the base of the ancestral rim at that time. Remember that the rim has been retreating constantly northeastward for at least 35 million years, leaving us a few clues for its various locations. This 6.4 mya-old trace of its location can be seen in the inverted topography (ridge of basalt) southeast of the Village of Oak Creek in the south wall of Jacks Canyon. Again, by measuring the distance from the inverted topography to the present rim (about two miles) and dividing by the number of years, we can calculate an average rate of retreat. Incredibly, this rate is about one foot every 606 years, almost identical to that calculated from the House Mountain trace of the ancestral rim. We can begin to feel confident now that we are learning something about how and when the rim evolved.

The Verde Fault, which played a major role in creating the Verde Valley as we know it today, also was active about eight million years ago. As much as 6,000 feet of rocks were downdropped to create the Verde Valley. Were it not for the location of this break in the earth's crust (which, incidentally, is

Sinagua ruin constructed with Late Paleozoic red rocks in a shelter of dark lava rocks. The Sinagua builders certainly must have paid attention to the quality and distribution of the various rock types in the Verde Valley. Photograph by Tom Brownold

where Jerome was built), the rich ores of the United Verde Mine might still be buried and beyond detection. It was the down faulting of the Verde Valley that fortuitously exposed these Precambrian ores.

The Verde Fault gave the Verde Valley its finishing touches, just as the retreat of the Mogollon Rim across the way to the east gave it its initial form. Continued and rapid subsidence of the valley floor caused the Verde River to become sluggish and, eventually, impounded. (Initially, geologists believed that lava from the Hackberry Mountain volcano, near Childs, had caused the river channel to become dammed, but recent studies on the age of these lavas suggests that they are too old.) With the river trapped in the basin, sediments began to accumulate on the valley floor. These sediments are known today as the Verde Formation. They are exposed everywhere in the central Verde Valley, at Montezuma Castle and Well, and near the towns of Camp Verde, Cottonwood, and Clarkdale. Interesting exposures can be observed along the road from Page Springs to Cornville, where the impounded sediments lapped onto the House Mountain volcano and eventually buried it completely.

The Verde Formation is over 3,000 feet thick in the central portions of the basin. It is composed of white limestone, brown mudstone, salt, and other evaporites. It accumulated for a six-million-year period until about two million years ago. Some of the "ramp basalt" flowed into this shallow lake basin, and these lava flows provide us with good age determinations for the deposits. Two of these interbedded flows are observed easily from the Watchtower Rest Area on Interstate 17. These have been dated at about 5.5 million years ago and originally were thought to have been derived from House Mountain. It is only since 1987 that we have known that House Mountain is three times older than this. Our knowledge is expanding rapidly.

The climate of the Verde Valley during deposition of the Verde Formation was dry, and the basin contained a very shallow lake, marsh, and playa environment. This is suggested by the caliche and salt deposits within the Verde Formation. It also tells us that the valley floor occasionally dried up. The Sinagua, a prehistoric people who lived near the present site of Camp Verde, constructed a salt mine (complete with supporting timbers) in the basin. This is one of the few localities in North America where native people used this type of technology.

Although rarely seen, the fossil evidence from the Verde Formation gives us a glimpse of life within the valley before the Ice Ages. It is a scene reminiscent of the grassy savannahs of east Africa because large mammals, now extinct, once roamed this place. At the Phoenix Cement Plant in Clarkdale, the giant tusks and jaws of a stegomastodon were unearthed in 1981. At another locality,

Ten-foot tusks of a Stegomastedon were unearthed at the Phoenix Cement Plant in Clarkdale by Professor Larry Agenbroad (with hat) in 1981. Photograph by Dale Nations

the footprints of one of these large elephants were found in a shallow water limestone, recording a single day almost five million years ago when this individual searched the shoreline for something to eat. Elsewhere, there is evidence for the presence of other large mammals—camel, antelope, tapir, and cats, many of which became extinct on our continent only 10,000 to 15,000 years ago.

By approximately two million years ago, the Verde basin was full of sediment. The river flowed first on top of the white limestones and then began to cut into them. House Mountain, now at an elevation of 5,127 feet above sea level, was uncovered (exhumed) from its sedimentary veneer, as were the various traces of the ancestral Mogollon Rim. Once the now familiar modern drainages of Sycamore, Oak Creek, Dry Beaver, Wet Beaver, and Fossil creeks cut their canyons into the edge of the Mogollon Rim, sediments from them were deposited throughout the area. One of the largest of these deposits is on Table Top Mountain, or Airport Mesa, in Sedona. Here, we find 50 to 75 feet of cobbles, gravel, and mud, identical in composition to that found today in Oak Creek.

There are other similarities which suggest that Oak Creek once flowed on top of Airport Mesa. The mesa has a gradient to the southwest of eighty feet per mile; Oak Creek is 70 feet per mile in the same direction. Both features seem to exit from the mouth of the Oak Creek Canyon. Both are much longer than they are wide. The only difference is that the deposits on Table Top Mountain are 700 feet above Oak Creek today. It is evident that at some time prior to the present, Oak Creek flowed on a surface much higher than it is now.

If we knew the exact age of these gravels, we could determine with precision how fast Oak Creek is cutting into the valley floor. This type of sediment, unfortunately, does not preserve fossils well, and we have not found specimens to date. We can only say that the youngest Verde Formation sediments are about two million years old, and they certainly once covered the area where Sedona is today. These Verde deposits had to have been eroded away before the gravels on Table Top Mountain were laid down. A good estimate of their age may be 1.5 million years ago. If correct, this means that Oak Creek is cutting into the Verde Valley floor at an average rate of about one foot every 2,150 years.

Verde Formation within the walls of Montezuma Well. The well was created when a cavern collapsed and spring water covered its floor. Photograph by Christa Sadler

CONCLUSION

The endangered Verde River is the primary geological agent to have shaped the modern Verde Valley. Will man replace the river as the main agent of change? Photograph by Ralph Hopkins

Volcanoes erupting on a primeval ocean floor, tropical coral seas, sand dunes blowing along the coastline of an ancient supercontinent, and a volcano spewing lava beneath a cliff of red sandstone—these are some of the colorful scenes that geologists have reconstructed from the rocks contained within the Verde Valley. Given the incredible forces of erosion in this area, it is astonishing that we are able to recall anything at all from our early heritage. And, yet, each succeeding generation of earth scientists has been able to bring back more of the ancient landscapes of the Verde Valley. This is the excitement of geology—to be able to resurrect the past. The rock record is the ultimate time machine that reveals to us the genesis of all life.

In the beginning, it was the economic rewards contained within the rocks of the Verde Valley that drew our attention to them. Increasingly, in the future, it may be the *intellectual* and *spiritual* rewards that the rocks can give which will satisfy our needs.

As we consider the vastness of time and the endless vistas in the area, there is one geological feature that captures our attention in the present—the Verde River. It is the river that has given us the soil, the river that gives us our crops, and the river that quenches our thirst in the desert. As more people come to this valley seeking a piece of "old Arizona," the demands placed on the river will threaten its very existence. Incredibly, man may become the primary agent of geological change in the Verde Valley. (Remember that the lush growth of cottonwood and sycamore, which still graces the floor of the Verde Valley, used to exist in the Salt River Valley where the city of Phoenix now is located.) By taking the time to learn of the valley's past and to understand the agents of change, we can limit how heavily the hand of man will fall on this place.

This volume is dedicated to the preservation of the Verde River as the vibrant, living heart of the Verde Valley. If the river survives and prospers, so may we.

ABOUT THE AUTHOR

Wayne Ranney is a geologist and freelance educator who specializes in making complex geological ideas accessible and understandable to interested nonspecialists. An instructor at Yavapai College in Prescott, Arizona, he frequently works as a lecturer with SPECIAL EXPEDITIONS on their cruise ships throughout the world, as a guide and boatman with WORLDWIDE EXPLORATIONS on southwestern rivers, and assists the Museum of Northern Arizona in docent training and the Ventures program.

Ranney received his Bachelors and Masters degrees from Northern Arizona University in Flagstaff, where his thesis work involved making a geologic map of the House Mountain volcano near Sedona. As a result of this research, he was awarded the Best Student Paper Award at the Arizona-Nevada Academy of Science Meeting in Tucson in April 1988, and the Edwin D. McKee Award at the MNA Geology Symposium, September 1988.

Stream cobbles in Sycamore Canyon.
Photograph by Stephen Trimble

SUGGESTED READING

Aitchison, Stewart. 1989. *A Guide to Exploring Oak Creek and the Sedona Area.* 111 pp. Contains descriptions and maps to the trails in the area and tells where to find interesting geological features.

Nations, J. Dale, Clay M. Conway, and Gordon A. Swann, eds. 1986. *Geology of Central and Northern Arizona.* Field Trip Guidebook for the Geological Society of America, Rocky Mountain Section Meeting in Flagstaff. Fascinating articles on Jerome ore genesis and Cenozoic evolution of the Mogollon Rim. 176 pp.

Ranney, Wayne. 1988. "Geologic History of the House Mountain area, Yavapai County." Unpublished Master's Thesis, Northern Arizona University, Flagstaff. Details the geology of a 75-square-mile area centered around the House Mountain Volcano. Map. Located in N.A.U. and M.N.A. libraries.

Smiley, Terah L., J. Dale Nations, Troy L. Péwé, and John P. Schafer, eds., 1984. *Landscapes of Arizona: The Geological Story.* Contains excellent essays and photographs on the origin of the landscape in Arizona. 506 pp.

Twenter, F.R. and D.G. Metzger. 1963. "Geology and Groundwater in the Verde Valley—The Mogollon Rim Region, Arizona." *U.S.G.S. Bulletin 1177.* One of the best of the early works on Verde Valley geology. Some nomenclature out of date, but otherwise very good. 132 pp.

Young, Herbert V. 1964. *Ghosts of Cleopatra Hill: Men and Legends of Old Jerome.* Jerome Historical Society. Colorful description of the old mining days in Jerome. Lots of old photographs. 183 pp.

Plateau Managing Editor: Diana Clark Lubick
Graphic Design by Dianne Moen Zahnle
Color Separations by American Color
Printing by Land O'Sun Printers
Typography by MacTypeNet